Food 79

边伐木边造林

Cutting Wood While Growing Forests

Gunter Pauli

[比]冈特·鲍利 著

[哥伦]凯瑟琳娜·巴赫 绘

李欢欢 牛玲娟 译

上海遠東出版社

特别感谢以下热心人士对童书工作的支持：

匡志强　宋小华　解　东　厉　云　李　婧　庞英元
李　阳　刘　丹　冯家宝　熊彩虹　罗淑怡　旷　婉
杨　荣　刘学振　何圣霖　廖清州　谭燕宁　王　征
李　杰　韦小宏　欧　亮　陈强林　陈　果　寿颖慧
罗　佳　傅　俊　白永喆　戴　虹

目录

Contents

猩猩忧心忡忡，加里曼丹岛的森林正不断被砍伐。也许很快他就没有可以悬挂的树枝了。这时他看见一只侏儒象正沿着河岸试图逃跑。

“你为什么这么匆忙呢？”猩猩问道。

The orangutang is worried. Forests are being cut down in Kalimantan. Soon he may not have a branch to hang from. He spots a pygmy elephant trying to escape along the river.

"Why are you in such a hurry?" asks the orangutang.

你为什么这么匆忙呢？

Why are you in such a hurry?

我们的家园快要消失了

Our home is about to disappear

“我们的家园快要消失了，这片物种丰富的茂密森林被棕榈树取代了，没人愿意住这里了。”侏儒象尖叫着说。

“我不清楚人类是否知道野生生物无法生活在非洲棕榈树种植园，但是棕榈树适合制造不会污染欧洲河流的肥皂。”

“Our home is about to disappear, the thick and diverse forest is replaced with oil palm trees where no one wants to live,” squeals the elephant.

“I wonder if people know that wildlife cannot live in African oil palm plantations. On the other hand, the palm oil is good for making soap that do not pollute rivers in Europe.”

“让欧洲的河流保持干净是很不错的想法，但要以破坏我们的家园为代价吗？这不合理。”

“唉，以前欧洲的河流被泡沫覆盖。后来，有人发现棕榈油制成的肥皂清洁度高，降解迅速，河面就不再有泡沫了。”

“这给了企业家在亚洲种植非洲棕榈树，破坏我们栖息地的特权吗？”

“Cleaning up rivers in Europe is a great idea, but at the cost of destroying our home? That can never be justified.”

“Well, Europe used to have rivers which were covered in foam. Then they discovered that palm oil-based soap cleans well, disappears fast, and keeps the rivers foam free.”

“Does that give these entrepreneurs license to plant an African tree in Asia and to destroy our habitat?”

棕榈油制成的肥皂清洁度高

Palm oil-based soap cleans well

他们最初并不知道

But at the beginning they didn't know

“不是，他们最初并不知道。”

“你是说，他们不知道种植非洲的树种会破坏我们家族世世代代生活的森林？”

“我相信他们不知道。但由于肥皂十分好用，越来越多的森林被破坏了，由棕榈树种植园取而代之。”

“No, but at the beginning they didn’t know.”

“You mean they didn’t know that African trees were going to be planted on destroyed forest land where my family has been roaming for ages?”

“I trust they didn’t know. But as the soap was a success, more and more forests needed to be destroyed to be replaced by plantations.”

“他们意识到这一点时，做了什么吗？”

“他们继续种植棕榈树。”

“什么？这是犯罪！”

“他们中有些人意识到了自己所犯的错误，为此感到难过，希望创造出可持续发展的棕榈林。”

“Once they had figured it out, what did they do?”

“They continued to plant the palm trees.”

“What? That is criminal!”

“Some of them realised their error, felt bad, and wanted to create sustainable palm oil plantations.”

创造出可持续发展的棕榈林

To create sustainable palm oil plantations

改正过去的错误

To reverse the mistakes of the past

“破坏土地来进行种植，怎么能称为可持续？更糟糕的是，这些棕榈林完全是空荡荡的，它们已经被破坏了，不会再有生命存在了！人类没有意识到吗？”

“巨大的破坏已经造成，但人类已经学会了如何改正过去的错误。”

“怎么做呢？”

“How can you call anything sustainable when it is planted on destroyed land? And what’s even worse is that these palm plantations are as good as empty forests. This damage is done and there will be no life here ever again! Don’t people realise that?”

“Great damage has been done, but we have learned how to reverse the mistakes of the past.”

“How?”

“人类在荒芜的大草原上种植了成千上万的松树，那里过去是一大片森林。”

“只种植松树，真是个坏主意。”

“松树只是开始，它是先锋植物。”

“那最后会发生什么？”

“People planted millions of pines in the barren savannah that was once a great forest.”

“Planting only pines is a bad idea.”

“The pine is only the beginning. It is the pioneer plant.”

“And what will happen in the end?”

在大草原上种植了成千上万的松树

Planted millions of pines in the savannah

出现了一片幼小的森林

A young forest can emerge

“也许你无法预测到最终结果，但你可以决定发展方向。”

“通往什么方向？我可以和我的朋友一起走一趟吗？”

“由于松树的绿荫遮挡，土壤温度很低。雨水渗过土壤，形成水源，帮助休眠种子生长，所以出现了一片幼小的森林。”

“You can never predict the end, but you can decide on the direction.”

“What direction is this road leading to? Can I walk it with my friends?”

“Thanks to the shade of the pine trees, the soil is now cooler and rainwater filters through it, producing drinking water, and helping all the dormant seeds grow so that a young forest can emerge.”

“太神奇了！”

“人们从每十株松树中移走八株，这样当地自然生长的植物就能够茁壮成长。于是当地的物种更丰富了。”

“这意味着我们可以边伐木边造林。我们都应该这么做！”

……这仅仅是开始！……

“Fantastic!”

“The community removes eight out of every ten young pine trees so that native plants that grow there naturally can thrive. This improves the biodiversity of the area.”

“That means we can cut wood while growing new forests. We should all do that!”

... AND IT HAS ONLY JUST BEGUN!...

……这仅仅是开始！……

... AND IT HAS ONLY JUST BEGUN! ...

Did You Know ?

你知道吗？

In three countries alone (Indonesia, Malaysia, and Papua New Guinea) 3.5 million hectares of forestland were lost to palm plantations between 1990 and 2010.

1990 年至 2010 年之间，印度尼西亚、马来西亚和巴布亚新几内亚三个国家就消失了 350 万公顷森林，用于种植棕榈树。

The loss of rainforests threatens the survival of the last remaining Sumatran tigers, the orangutang, and the pigmy elephant, as well as the world's largest biodiversity of vascular plants.

热带雨林的减少威胁着濒临灭绝的苏门答腊虎、猩猩和侏儒象的生存，以及世界上最多样化的维管束植物。

When deforestation occurs on peat soils, enormous amounts of CO_2 is emitted. It is not necessary to deforest so much land as 12.5 million hectares of degraded land is already available for tree plantations.

在泥炭土上滥伐森林，会释放大量的二氧化碳。有 1 250 万公顷的退化土地可以用来造林，没有必要如此滥伐。

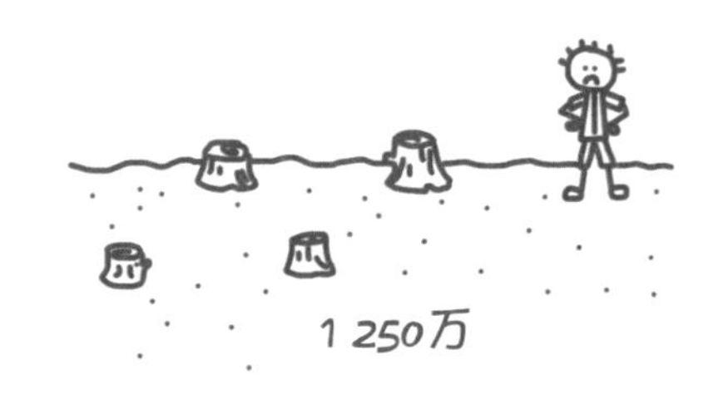

Palm oil is the most used vegetable oil in the world. 42 million tonnes of palm oil are exported each year, representing 65% of all internationally traded oil. It is used in everything from chocolate to toothpaste, to lipstick, and cookies.

棕榈油是世界上使用最多的植物油，每年出口 4 200 万吨，占全球油量交易的 65%。它广泛用于制作巧克力、牙膏、唇膏和小甜点。

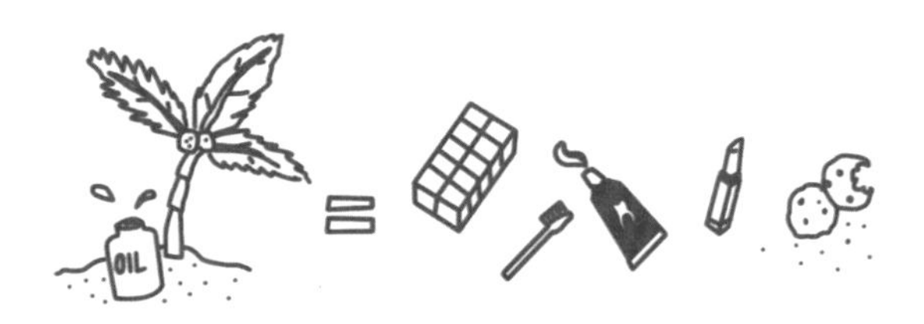

Deforested land in the Vichada Department of Colombia regenerated its rainforest more than 200 years after its destruction. Las Gaviotas demonstrated that the combination of science and human commitment could correct the errors of the past.

在热带雨林遭破坏的 200 年后，哥伦比亚比查达省在原来的土地上重建了一个雨林。拉斯卡维塔斯的实践证明，科学加上人类的努力，可以改正以前的错误。

The regeneration of the rainforests in Las Gaviotas includes blending oil palm trees into the new forest. The oil is processed locally and made available for local consumption. The excess wood is used for energy and to make fuel.

拉斯卡维塔斯雨林的重建包括在新森林里混合种植棕榈树。棕榈油在当地生产，供当地消费。木材用于能源和燃料生产。

The denuded land in Colombia had 17 plants, including 11 non-native grasses. 25 years after the initiation of the project, the regenerated forests have 256 plant species.

哥伦比亚这片被砍伐过森林的土地曾经只有 17 种植物，包括 11 种外来物种。项目启动 25 年后，重建的森林中已经有了 256 种植物。

Pollen studies enables the identification of all the plants that once thrived in this region for over six thousand years. Small bio-reserves along the rivers provided the natural source of biota carried by birds, bees, and the wind to “fill” the forest.

通过花粉研究，确定了过去 6 000 年曾在这片土地生长的所有植物。沿河的小型生物保护区是这片区域的生物源头，鸟类、蜜蜂和风把这些生物带入森林。

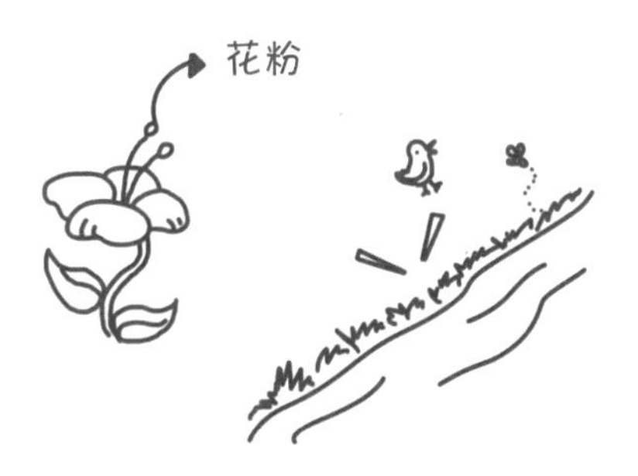

Think about It
想一想

Would you like to eat chocolate knowing that the oil it is made of is responsible for destroying rainforests and the habitat of tigers, elephants, and primates?

喜欢吃巧克力吗？你知道用来做巧克力的棕榈油是毁坏热带雨林，破坏老虎、大象和猩猩栖息地的原因吗？

如果你不知道自己的行为有不良后果，那就不是故意的，可以被原谅。但是一旦知道了后果，你会怎么做？

If you did not know your actions had negative consequences, they were unintended and forgivable, but what do you do once you find out the truth?

Do you think that planting one tree a year is enough? How many trees are needed to grow a forest?

你认为一年种一棵树够吗？需要种植多少棵树才能形成一片森林？

如果你有办法能消除已经造成的伤害，你还会对自己之前犯的错误感到难过吗？

Would you still feel bad about your errors of the past if you knew there was a way to reverse the damage done?

Check all the food products in your house. Which ones contain palm oil? Most of the time the label only states “vegetable oil”, without any further details. Put all the products you find together on a table and take a photograph of it. Then you will realise that palm oil has become a big part of our lives.

检查家里所有的食品，哪些含有棕榈油？大部分标签只标有“植物油”，没有更详细的说明。把你找到的所有食品放在一张桌子上，拍一张照片，然后你会发现棕榈油已成为我们生活中的重要组成部分。

学科知识

Academic Knowledge

生物学	非洲象、亚洲象和侏儒象的区别；濒临灭绝的苏门答腊虎；具有丰富生物多样性的沿河森林走廊的重要性；森林重建时树木很快就开始锁住土壤，生长过程中，多达80%的树会被移除，以便让最强壮的树生长；重建森林过程中，热带松树在贫瘠草原恶劣环境中的生存能力。
化　学	棕榈油的饱和脂肪酸含量低，具有抗氧化性，含有脂肪酸、甘油和胡萝卜素；棕榈油用于生产甲酯和加氢脱氧生物柴油。
物　理	植物遮挡阳光，树荫降低了土壤温度，使其低于雨水温度，能够更好地吸收水分，保护种子和幼苗免受大量紫外线照射。
工程学	棕榈油生产需要研磨、分馏精炼、结晶、固体分离、熔化、脱胶、过滤和漂白；红棕榈油冷榨装瓶，用于烹饪；制油产生的有机废物能制成燃料。
经济学	一个标有“有机”或“可持续”的产品能轻易取代同类或稍有不同的产品；许可经营一项业务似乎还包含允许破坏和污染。
伦理学	棕榈油曾是非洲和亚洲穷人的主要食品，使用于成千上万种产品，并转换成燃料，这使棕榈油价格提升，引起了棕榈油究竟是用于燃料还是食品的争论；如果是无意导致的后果，那么有责任改正以前的错误。
历　史	棕榈油的使用最远可追溯到5 000年前的埃及；19世纪70年代，棕榈油是加纳和尼日利亚的主要贸易产品；英国工业革命期间，棕榈油是优选润滑剂。
地　理	加里曼丹地区隶属于印度尼西亚；棕榈树生长在全球的热带地区；雨水丰富的森林一旦消失，就会出现稀树草原，后者之后会变成沙漠。
数　学	每公顷棕榈树每年生产3.7吨棕榈油，是大豆产量的10倍、油菜籽产量的5倍；全球种植了将近1亿公顷大豆、1 000万公顷棕榈树，但两者都被少数企业垄断。
生活方式	消费者还未意识到时，棕榈油就已经取代了多种当地油产品；消费者只看见产品及其功能，不会意识到自己的生活会影响到那些遥远的国家。
社会学	西方文化的线性时间轴观念：那些已发生的事无法改变；东方文化的循环时间轴观念：丢失的机会会回来，因此已发生的错误能改正。
心理学	“格式塔”心理学强调看到整体以及在对某事形成看法前了解事情的来龙去脉；当我们意识到能挽回损失，过去的负面影响将来能被正面影响取代时，会觉得安心；如何把愤怒转变成兴奋的能量。
系统论	单一种植（如在一片土地上只种棕榈树、玉米、小麦或西红柿）常常会导致很多意想不到的问题，企业很少考虑这些问题，因为它们只注重核心业务和利益，忽略了潜在的外部影响。

情感智慧

Emotional Intelligence

猩　猩

猩猩关心森林里的朋友，表现了同情心。他仔细思考并意识到使用棕榈油的利弊，认为快速生物降解的棕榈油能清洁欧洲河流。猩猩认为企业没有意识到自己造成的破坏，而且在发现事实后也没有采取任何措施。猩猩肯定了那些对所犯错误感到难过，进而想要解决问题并促进可持续种植的人，他坚定地相信过去的负面影响能消除。在侏儒象的敦促下，猩猩给出了一个积极解决方法的例子：从松树开始种植一片新森林。他解释了已经做的尝试，也承认现在还没有明确结果。但他向侏儒象保证，这一发展方向是前进之路。

侏　儒　象

侏儒象容易激动，说话语调高昂。他担心失去自己的家园，基于道德和正义的标准，表明毁坏家园永远都没有正当理由。他说人类不应该为了获得利益而去做一些坏事。他质疑颁发企业运营许可证的系统，起初那些企业没有意识到会造成破坏，但发现事实后还继续之前的行为，侏儒象为此感到很生气。他作出肯定的判断，认为这是犯罪，并对人们执意认为在退化的土地上进行新型种植是可持续种植提出疑问。当猩猩指出以松树为先锋植物重建森林的方法时，侏儒象起初强烈反对，但猩猩讲清了过程后，他同意了这个方法，由愤怒变成了兴奋。

艺术

The Arts

我们看到了森林被砍伐后土地被焚烧、树木被移走的图片，也看到了美丽的热带雨林图片。为什么不画两幅图，一个展现翠绿，另一个展现褐色和淡绿？讨论一下这种改变的原因，讲述热带雨林被毁变成单一种植的棕榈林的情况，以及改正错误的可能性：有了决心和耐心，光秃秃的土地也能重新变成雨林。

思维拓展

Systems: Making the Connections

在寻找高产作物的过程中，人类发现了起源于非洲西部的棕榈树，将它纳入农业产业化的一部分。这种棕榈树的产油量是大豆的10倍，且相当稳定。于是许多动物油和植物油被棕榈油取代，衍生出了一个大型产业，集中于马来西亚和印度尼西亚。不幸的是，为了生产更多标准化的油，人类扩大棕榈树种植，导致大量本地物种死亡。人类面临的主要挑战是没有足够的土地来种植棕榈树，因此只能牺牲雨林、牺牲老虎、猩猩和大象的栖息地。如今，大量的破坏仍在继续。在热带地区，单一种植棕榈树导致真菌病害，从内部破坏棕榈树。而大量使用杀真菌剂伤害了生物网里的所有其他成员，造成了光秃或贫瘠的森林。过去被烧掉的木材，现在越来越多地用作燃料和家具制造，但问题依然存在。在马来西亚，最初的棕榈园越来越多地变成了房地产项目。快速的城市化进程和住房的高需求导致森林有了第二个被砍伐的理由。吉隆坡的城市化模式证实了农田变房产，土地价值会增长1 000倍。亚洲还将要建立至少1 000个百万人口城市，土地会像水、能源和公路一样，变得非常值钱。然而，也还有另外一个选择，即重建森林。我们不能继续在世界各地砍伐森林，改变大气的化学成分了。因为过去是我们砍伐了森林，利用了土壤，所以现在重要的是，我们应保持积极的态度，并承担重建森林的所有责任。修复表层土壤，让重生的生态系统创造价值，提供食物、水、工作和收入——这些都是社会的迫切需求。

动手能力

Capacity to Implement

无论你生活在世界何处，都一定会与某种植物油或者动物油接触。让我们研究一下可选择的本地油，而不是全球销售的油。它没有被标准化，但会让我们发现哪些油在批量生产中被遗弃。你能列出本地的油吗?

故事灵感来自

This Fable Is Inspired by

更家悠介

Yusuke Saraya

更家悠介出生于日本大阪，获得微生物学学位，大半生就职于更家有限公司，现任总裁。悠介是日本青年商会主席，负责开拓国际关系。作为肥皂的主要生厂商，他知道棕榈油对环境的挑战，决定参加可持续发展棕榈油圆桌会议，并成立了一个基金会，救助加里曼丹侏儒象。他还带领自己公司的研发团队找到棕榈油的一系列替代品。他投资的新产品，不仅清洁度高，而且确保人类和地球的健康。他也是联合国大学第十届和第二十届零排放国际大会主办人。

图书在版编目（CIP）数据

冈特生态童书.第三辑修订版：全36册：汉英对照 / （比）冈特·鲍利著；（哥伦）凯瑟琳娜·巴赫绘；何家振等译.—上海：上海远东出版社，2022
书名原文：Gunter's Fables
ISBN 978-7-5476-1850-9

Ⅰ.①冈… Ⅱ.①冈… ②凯… ③何… Ⅲ.①生态环境-环境保护-儿童读物—汉、英 Ⅳ.①X171.1-49

中国版本图书馆CIP数据核字（2022）第163904号

著作权合同登记号图字09-2022-0637号

策　　划　张　蓉
责任编辑　程云琦
封面设计　魏　来李　廉

冈特生态童书
边伐木边造林
[比]冈特·鲍利　著
[哥伦]凯瑟琳娜·巴赫　绘
李欢欢　牛玲娟　译